# Amethyst Gemstones

A Collection of Historical Articles on the Origins, Structure and Properties of Quartz

By

Various Authors

Copyright © 2011 Read Books Ltd.
This book is copyright and may not be
reproduced or copied in any way without
the express permission of the publisher in writing

**British Library Cataloguing-in-Publication Data**
A catalogue record for this book is available from
the British Library

# Contents

The Natural History of Gems – Or, Decorative Stones.
C W King..................................................................*page* 1

A Manual of Precious Stones and Antique Gems.
Hodder M Westropp.....................................................*page* 11

Questions Answered on Gems and Jewellery.
Kathleen Gough...........................................................*page* 15

Diamonds and Precious Stones - Their History, Value and Distinguishing Characteristics. Harry Emanuel................*page* 17

A Hand-Book of Precious Stones. M D Rothschild..............*page* 21

An Introduction to Crystallography. F C Phillips................*page* 23

# AMETHYSTUS: 'Αμέθυσος: 'Αμέθυστος: *Amethyst.*

OUR Common Amethyst, and the stone (perhaps) generally designated amongst the ancients by this name, is nothing more than rock crystal coloured purple by manganese and iron, and on this account is more properly termed in modern mineralogy Amethystine Quartz. It is therefore of an entirely distinct species from the true Oriental Amethyst, a most rare and valuable variety of the Precious Corundum, and which is in fact a purple Sapphire, but its purple shows little of the red (*ponceau*) seen in the common Amethyst, being rather an extremely deep shade of violet.\*
The name of "Oriental" is, however, improperly applied by the English lapidaries to the Amethystine Quartz when very brilliant and of two shades of colour (qualities distinguishing the Indian from the German), the true gem of the name, from its rarity, being known to very few among them.

The name "Amethyst," though most probably a mere corruption of the Eastern name for the stone, a trace of which seems preserved in the Hebrew Achlamath,† was by the fanciful Greeks interpreted as though formed in their own language, from ἀ μεθὺ "wineless," and on the strength

\* The common Amethyst, formerly brought from Carthagena in Spain, and now only to be met with in old-fashioned pieces of jewelry, alone of its species exhibits this pure violet colour.

† Perhaps the true origin is, as Von Hammer suggests, the Persian "Shemest."

of this etymology the gem was invested by them with the virtue of acting as an antidote to the effects of wine.* Hence the point of several epigrams in the Anthology, as that of Antipater's (or Asclepiades) on the signet of Cleopatra, an Amethyst engraved with the figure of Μέθη, the genius of intoxication (ix. 752)—

> Εἰμὶ Μέθη, τὸ δὲ γλύμμα σοφῆς χερὸς ἐν δ' ἀμεθύστῳ
> γέγλυμμαι· τέχνης, δ' ἡ λίθος ἀλλοτρίη·
> ἀλλὰ Κλεοπάτρας ἱερὸν κτέαρ· ἐν γὰο ἀνάσσης
> χειρὶ θεὸν νήφειν καὶ μεθύουσαν ἔδει.

> "A Mænad wild, on amethyst I stand,
> The engraving truly of a skilful hand;
> A subject foreign to the sober stone,
> But Cleopatra claims it for her own;
> And hallow'd by her touch, the nymph so free
> Must quit her drunken mood, and sober be."

Another, more briefly playing on the same fancy (ix. 748)—

> Ἀ λίθος ἐστ' ἀμέθυστος ἐγὼ δ' ὁ πότας Διόνυσος,
> ἢ νήφειν πείσει μ', ἢ μαθέτω μεθύειν.

> On wineless gem, I, toper Bacchus, reign;
> Learn, stone, to drink, or teach me to abstain!

Or, as Pliny explains the import of the *name* (xxxvii. 40), "because these gems never come up to the colour of wine, since before they touch it their lustre falls off into the colour of the *viola*" (*i.e.* pink cyclamen).

Pliny divides the Amethystus into five kinds, the Indian holding the first rank; others coming from Arabia Petræa, Armenia Minor, Egypt, and Galatia; inferior sorts from Thasos and Cyprus. The Indian displayed the precise colour of the imperial purple; a variety of these "degenerated into that of the Hyacinthus (Sapphire), and was

---

* Mohammed Ben Mansur affirms that wine drunk out of an amethyst cup does not intoxicate.

called by the Indians Sacondion, Sacon being their term for that particular tint; if still lighter, it took the name of Sapenos." The fourth sort was of a wine (we should say, "Burgundy") colour; the fifth and worst of all was so pale as to resemble Crystal. The most admired tint was where a slightly rosy hue shone out from amidst the purple, and became more conspicuous when viewed by transmitted light (in suspectu); such were distinguished by the title of Pæderotes (Cupids), or the "Gems of Venus," on account of the pre-eminence of their kind and their beauty.

The deeper the tint the less brilliant is the stone, for which reason the ancient engravers preferred the light-coloured variety, which of all gems, next to the Jacinth, possesses the greatest degree of lustre; or they may have used it for cutting upon on account of its greater cheapness, remarked above. That Amethysts\* of a fine colour (now so worthless) were deemed too valuable by the ancients to have their substance diminished by the sinking of intagli into them, appears from many examples extant. They were either worn as mere ornamental jewels uncut, or else polished to an extremely convex form, presenting in their exact centre a diminutive intaglio, a Gorgon's Head, or a mask, in the nature of a talisman that augmented the supposed virtue without detracting much from the native beauty of the gem. Pliny notes the suitableness of all the Amethyst family for engraving upon (scalpturis faciles), a sufficient proof that no species of this stone was the Hyacinthus (the common explanation of archæologists from De Boot to K. O. Müller), which Solinus with justice calls

---

\* The rich Indian Amethyst evidently was then equally precious with the Sapphire; Pliny undeniably regarded the latter as merely a variety of it; for this reason the two are often found set side by side in ancient jewels.

the hardest of all gems, and only to be touched by the Diamond-point.

Intagli of all dates and in every style occur upon Amethysts, but so much more generally on the pale sort that an engraving upon one of a rich dark colour may, on that very ground, be suspected as modern. Besides the foregoing remarks as to the high value of such a shade in antiquity, the modern artists have usually employed the Hungarian Amethyst as being now the most abundant, of which the tint is a fine reddish purple, though the lustre is far below that of the Indian. Amongst the few exceptions to this rule that have come under my own notice is a head styled of Mithridates, but, in my judgment of some Bactrian king of the line of Euthydemus; perhaps the noblest Greek portrait in existence, cut in a large Amethyst of the deepest violet colour, found a century ago in India; which, however, being doubtless the royal signet, rather corroborates than weakens the previous statement. For such a use the most precious material procurable would naturally have been selected; "ut alibi ars, alibi materia esset in pretio," to use Pliny's expression. Another fine Greek intaglio, a head of Pan, in front face, on a similar stone, the antiquity of which could not be called in question, was in the Uzielli Cabinet; and, above all, stands the unrivalled Marlborough Omphale, the first amongst the numerous repetitions of that favourite subject, where the Amethyst (of the Indian kind) possesses equal lustre and richness of colour. In a large circular, convex stone of this sort is engraved the Berlin Atalanta, justly styled by Winckelmann "une des gravures les plus parfaites qu'on puisse voir." The swift-foot nymph is figured running at full speed and holding down with both hands the folds of her voluminous peplum distended by the agitated air. Through its gauzy material

## AMETHYSTUS.

the elegant contour of her whole body is distinctly visible. She is turning round her neck and looking back as if about to stop to pick up the golden apple thrown down by her competitor.

Heads, and even busts, both in full and in half-relief, often occur of antique workmanship in this stone: as some perfectly preserved remains show they served to complete statuettes in the precious metals. The grandest of Medusa-heads, the Blacas, is carved out of an Amethyst of the darkest violet, two inches in diameter. Although the Amethyst came into use amongst the earliest materials used by the gem-engraver, for we find in it an abundance of Egyptian charms (pendants for necklaces), in the form of vases, shells, hands, &c., and sometimes scarabæi, the last of Etruscan work also, and Roman intagli in it are sufficiently numerous, yet it is a singular fact that we rarely meet with works in the highest style executed in this material. Probably the superior kind was too precious to be so employed, whilst the paleness of the other and cheaper sorts was repugnant to the taste of first-rate artists.

But besides the stone known at present by the same name, there can be little doubt the Roman "Amethystus" included amongst its varieties a totally distinct species of gem—some kinds of our common garnets. This conjecture is supported by the authority of De Boot, who says (ii. 30), "Amethystus veterum nunc Granati nomen obtinet."*

---

* There can be no question that the Amethyst (purple quartz) of the moderns was considered by the ancients as one species of their Jaspis. In no other way can Pliny be understood (37), where, describing the latter, he thus classifies them: "Minus refert nationes quam bonitates distinguere. Optima quæ purpuræ aliquid habet; secunda quæ rosæ; tertia quæ smaragdi." And such a classification is perfectly accurate, all being equally quartz crystals, variously coloured by different metallic oxides. Again, he alludes to another *Jaspis* resembling the Sard;

We cannot resist such an inference if we carefully examine some of the characters given by Pliny of certain varieties of this gem. Thus he describes the Indian as "having the exact tint of the royal purple, and the dyers direct their endeavours to produce the colour, taking this gem for their pattern. For it diffuses a hue softly gentle to the sight, neither does it flash upon the eye like the Carbunculus." Be it remembered that he has already described the best purple as the colour of clotted blood, dark in one aspect, bright red if viewed against the light.* Again, we find "the fourth sort has the colour of wine;" now Italian wine generally (and more especially that grown about Rome) shows the richest Burgundy colour, than which nothing more accurately expresses the deep hue of the common Pyrope. It is a manifest absurdity to suppose a comparison between the bluish red of our Amethyst and the unmixed red of various shades peculiar to any sort of wine.† Again, his "Amethystus" was exactly counterfeited by staining amber with either alkanet-root, or murex-blood; both *reds* with no tinge of *blue*. The Carbunculus of Pliny was doubtless our Spinel Ruby, and to the eye alone (the sole criterion of the ancient lapidary) the Oriental (Siriam) Garnet and the Spinel are almost undistinguishable from

another mimics the viola (or pink cyclamen); the Cappadocian was a sky-blue mixed with purple (ex purpura cærulea), but dull and not lustrous. The last definition applies exactly to our German Amethyst.

* A description closely applying to what is now called the Carbuncle. This dark shade of the ancient Tyrian dye is well exemplified by the remark of Augustus, preserved by Macrobius amongst his other facetiæ. When that prince was finding fault with the darkness of some purple purchased to his order, and the vendor repeatedly bade him "hold it up higher and look at it," he retorted, "Must I then be always walking on a balcony if I wish the Romans to know I am richly attired?"

† Theophrastus includes (30) τὸ ἀμέθυσον in his list of ring-stones, adding that it is like wine (οἰνωπὸν) in colour.

## AMETHYSTUS.

each other, as I have had frequent occasion to observe in looking over examples of both species which have come down to us from Roman times with engravings upon them. Again, in no other manner is it possible to understand what kind of precious stone Heliodorus is describing as set in the king of Ethiopia's ring (Æth. v. 13). "And so saying, he put into his hands a ring, one of the royal jewels, an extraordinary and astonishing thing, the shank being formed of electrum, and the beasil flaming with an Ethiopian Amethyst, in size about the circumference of a maiden's eye, but in beauty far surpassing either the Iberian or the British sort. For the latter blushes with a feeble hue, and is like a rose just unfolding its leaves from out of the bud, and beginning to be tinged with red by the sunbeams. But in the Ethiopian Amethyst, out of its depth flames forth, like a torch, a pure and as it were a Spring-like beauty; and if you turn the stone about as you hold it, it shoots out a golden lustre, not dazzling the sight by its fierceness, but resplendent with cheerfulness. Moreover, a more genuine nature is inherent in the species than is possessed by any brought from the West, for it does not belie its appellation, but proves in reality to the wearer an antidote against intoxication, preserving him sober in the midst of drinking-bouts." This gem was engraved with a youthful shepherd and his flock, of which the tasteful bishop proceeds to give a pretty description, in which he again dwells upon the "golden" tints commingled with the flamy hues of the stone. The same interpretation must be put upon the more obscure language of the somewhat later Epiphanius in his 'Treatise on the XII. Stones of the Rationale,' where under IX. he gives "The stone Amethyst: this in proportion to its circumference (" excellence " must be the true reading) is of a deep flame colour, or sometimes paler, sending forth

## NATURAL HISTORY OF GEMS.

out of the inside a vinous appearance. Of it there are various species: one of the sorts is very similar to a clear hyacinthus (Sapphire), the other to the murex-blood, *i.e.* Tyrian dye. They are found in the mountains and on the coast of Libya." Again, Pliny's definition of the *Pæderos* suits no other gem so exactly as a particular variety of the Almandine sometimes met with amongst antique Garnets, the tint of which is truly roseate, not purple. One such in my possession is engraved with a Cupid proudly bearing off the spoils of the vanquished Hercules; there is good reason to suspect that the popular name of the gem had influenced the choice of the subject.

Throughout the Gothic period the common Amethyst held the same rank as it had enjoyed with the Roman jeweller, and continues to keep company with the Sapphire in the ornaments of the priest or prince. Even as late as the year 1600 a perfect Indian Amethyst is valued at half the price of the Sapphire, viz., one thaler or crown for the first carat. For higher weights De Boot gives a curious rule; to add together the weight, and value in thalers, of the stone preceding in his table the number of the one in question, and thus brings up the value of one of 20 carats to the high figure of 201 thalers.

Even in the last century this now despised stone was held in high estimation, when Queen Charlotte's necklace of well-matched Amethysts, the most perfect ever got together, was valued at 2000*l.*; at present it would not command as many shillings, so swamping has been the importation of late years of German Amethysts and Topazes (purple and yellow crystals of quartz), which are got in endless abundance from various parts of Hungary, Bohemia, Saxony, and notably at Oberstein, where they are cut and polished expeditiously and cheaply by water power (on sandstone wheels turned by the stream of the Nahe), and

## AMETHYSTUS.

despatched into all parts of Europe to be made up into cheap articles of jewelry. They are also found plentifully about Wicklow, in Ireland. Barbot mentions a crystal of Amethyst as recently brought to Paris of the astonishing weight of 65 kilos. (about 140 lbs.). When the gem was in fashion, it was formerly imported largely from the East Indies, and these were light coloured, the purple being shaded unequally, but extremely lustrous. The colour of the Amethyst can be dispelled by a careful roasting in hot ashes. Hence in the last century, when it was the great desideratum with the jewellers to obtain a suite of stones all exactly of the same tint, they were able to bring about this result by subjecting the several pieces to the heat for a greater or less time, until they were all reduced to the same shade of purple. According to modern usage this is the only gem it is allowable to wear in mourning.

The artists of the Renaissance eagerly availed themselves of these huge and beautiful crystals to carve them into those fanciful yet elegant vases, so acceptable to the taste of their age. The Parisian Collections offer the choicest specimens of their skill in this line. A cup, shaped as a shell, seven inches long and deep, by six wide, valued at 1800 francs; also an urn eight inches high, fluted and elaborately decorated with engravings, are enumerated in the former treasury of the Crown.

This stone is one of the earliest that figure in the list of talismans, or gems whose native virtues were heightened by the sigil engraved upon them, a superstition still in its infancy in the age of Pliny, when, although the medical virtues of many gems were generally admitted, the doctrine of their supernatural powers was as yet ridiculed by the learned as a figment of the credulous East. Thus under this head Pliny remarks that "the lying Magi hold out that these gems are an antidote to drunkenness, and take their name from this property.* Moreover, that if the names of the Moon or Sun be engraved upon them, and they be thus hung about the neck from the hair of a baboon, or the feathers of a swallow, they are a charm against witchcraft. They are also serviceable to persons having petitions to make to princes: they keep off hailstorms and flights of locusts with the assistance of a spell which these doctors teach." Where it may be remarked that the names of the Moon and Sun must have been the mystic names of these luminaries emblazoned in a sacred *Oriental* language, for the same in Greek or Latin would have been too simple to captivate the faith of their dupes, and besides, never are to be seen on any similar relics. Doubtless we have here an allusion to the ΙΑΩ and ΑΒΡΑΣΑΞ, words of that actual signification, and which in the following century appear on such countless multitudes of talismanic gems connected with Magian or with Egyptian superstitions.

* A notion curiously illustrated by a four-sided Amethyst pendant (Blacas) presenting on each face a Bacchante in a different attitude of frenzy: a fine Greek work.

## AMETHYSTUS.—AMETHYST.

Among stones of a purple colour, Pliny gives the first rank to the amethyst of India, a stone which is also found, he says, in the part of Arabia that adjoins Syria, and is known in Petra, as also in Lesser Armenia, Egypt, and Galatia; the very worst of all and the least valued being those of Pharos and Cyprus. Another variety approaches more nearly the hyacinthus (sapphire) in colour: the people of India call this tint *socon*, and the stone itself *socondion*. Another was in colour like that of wine, and a last variety but little valued, bordering very closely upon that of crystal, the purple gradually passing off into white. A fine amethyst should always have, when viewed sideways (*in suspectu*), and held up to the light, a certain purple effulgence, like that of carbunculus, slightly inclining to a tint of rose. To these stones the names of *pæderos* and 'Venus' eyelid' (*Veneris gena*, Ἀφροδίτης βλέφαρον) were given, being considered as particularly appropriate to the colour and general appearance of the gem.

The name which these stones bear, originates, it is said, in the peculiar tint of their brilliancy, which, after closely approaching the colour of wine, passes off into a violet, without being fully pronounced. "All these stones," Pliny adds, "are transparent, and of an agreeable violet colour, and are easy to engrave. Those of India have in perfection the very richest shades of purple."

At the present day the finest amethysts come from India, and lapidaries apply the term Oriental to the amethystine quartz when of a very brilliant violet tint, and of two shades of colour (qualities distinguishing the Indian from the German). This stone must be, however, carefully distinguished from the true Oriental amethyst, which is a sapphire of a violet colour.

"Intagli of all dates," Mr. King says, "and in every style, occur upon amethysts, but so much more generally on the pale sort that an engraving upon one of a rich dark colour, may, on that very ground, be suspected as modern. Although the amethyst came into use amongst the earliest materials used by the gem engraver, for we find in it an abundance of Egyptian charms (pendants for necklaces), in the form of vases, shells, bands, &c., and sometimes Scarabæi, the last of Etruscan work also, and Roman intagli in it are sufficiently numerous, yet it is a singular fact that we rarely meet with works in the highest style executed in this material. Probably the superior kind was too precious to be so employed, whilst the paleness of the other and cheaper sorts was repugnant to the taste of first-rate artists."[e]

Some fine Greek intagli occur in this stone. Mr. King mentions, among others, the Marlborough Omphale, on an amethyst (of the Indian kind) of superior lustre and richness of colour, and the Berlin Atalanta engraved on a large circular convex stone. Among other celebrated engravings

Diana of Appollonius.

Pallas of Eutyches.

Medusa.

in amethyst are the Pallas of Eutyches, deeply engraved on a pale amethyst, the Achilles Citharœdus of Pamphilus (Paris). The Diana, of Appollonius (Naples), the Medusa (Blacas), the Mecænas of Dioscorides (Paris), a head of Pan, deeply sunk in a pale amethyst inscribed ΣΚΥΛΑΞ, in the Blacas Collection. In the Devonshire Collection is a magnificent amethyst intaglio, bearing the bust of Shappur I.

[e] "Natural History of Gems," p. 31.

of the race of the Sassanides.⁷ This stone (1¼ × 1 inch oval) forms the centre in the comb⁸ belonging to the *parure* of antique gems, the property of the Duke of Devonshire. In the Florentine collection is a large amethyst with the portrait of Mithridates the Great. "Heads, and even busts," Mr. King writes, "both in full and in half relief, often occur of antique workmanship in this stone, as some perfectly-preserved remains show they served to complete statuettes in the precious metals. The grandest of Medusa heads, the Blacas, is carved out of an amethyst of the darkest violet, two inches in diameter."

According to some authorities, the name amethyst has been derived from α not, μεθύω to intoxicate, on account of its being a supposed preservative against inebriety. Von Hammer suggests the Persian *shemest* as the true origin of the word.

## AMETHYST

**124.—What is an amethyst?**

The amethyst, queen of the quartz gems, is a very lovely semi-precious stone varying in colour from pale violet to deep purple, and at one time it was fairly costly. But that was in the days when it was comparatively rare, and not found in most parts of the world as it is now.

**125.—Are there any special virtues attached to this stone?**

It is supposed to be a remedy against drunkenness. This belief originated with the ancient Greeks, and probably had something to do with its colour. In olden times drinking cups were made of amethyst and people with an alcoholic thirst were given these cups filled with water which then took on the colour of wine. Maybe they were fooled, but it is hardly likely.

**126.—Is this gem much in demand to-day?**

It is not as popular as it was in our grandmothers' days, but there is sure to be a to-morrow when the limelight of fashion is turned upon it again.

**127.—Should amethysts be set in gold or a white metal?**

A gold background is best, especially for very deep amethysts. Sizeable stones look perfect surrounded by pearls, and this style makes desirable brooches and " clumsy " rings, and, too, there are some beautiful gold-set amethyst necklaces to be found in shops which deal in old jewellery.

**128.—Can amethysts be engraved?**

These stones have always been popular with the engraver and were much in demand for intaglios.

## The Amethyst.

The next variety of the vitreous portion of the quartz family is called amethyst, which is of a fine violet colour, passing from white to a deep purple, sometimes

## Amethyst.

in the same specimen. The deep purple-coloured specimens are frequently called oriental, even by jewellers and lapidaries, although the oriental amethyst is an entirely different stone, which has already been described under the head of sapphire. The colour of this gem is by some supposed to be derived from a trace of the oxide of manganese. Later analyses, however, have discovered also silica, iron, and soda. Heintz obtained, from a very deep purple Brazilian amethyst,—

| | |
|---|---|
| Oxide of iron | 0·0187 |
| Lime | 0·6236 |
| Magnesia | 0·0133 |
| Soda | 0·0418 |

The amethyst is found in India, Ceylon, the Brazils, Persia, Siberia, Hungary, Saxony, Spain, etc. A fine vein is said to exist near Kerry, in Ireland. In Oberstein it is found in a trap rock, in geodes of agate. These geodes are sometimes as much as two feet in diameter, hollow, and filled with crystalized amethyst of a fine colour. Similar geodes are also said to exist in India.

This variety of quartz, in common with some other of the vitreous members of the family, possesses a peculiarly minute, wrinkled, or wavy fracture on the fresh-broken surface, resembling the impression of the skin of the thumb on a waxy substance. Sir David Brewster classes all kinds of quartz having this peculiarity as amethysts, without regard to their colour.

This gem is found in pieces of considerable size, and,

## Amethyst.

from its beautiful colour and play, is much used in jewellery. Many years ago, amethysts were of considerable value, ranking next to the sapphire, and when fine selling at 30s. per carat; but immense quantities having been sent from the Brazils, they became common, gradually went out of fashion, and became nearly valueless. Latterly, however, the taste for them has revived, and at the present time they are gaining ground in public estimation. A fine clear deep-coloured amethyst, of the size of a two-shilling piece, is worth from £10 to £15; smaller sizes and inferior qualities are sold at from 2s. to 100s.

The amethyst is cut in various ways; but the mode which best shows the beauty of the stone is the brilliant-cut with a rounding table,—that is to say, cut like a diamond, but with the table, or flat part of the stone, slightly domed. Very few amethysts are cut in this country, as the price of labour is too high: great quantities are sent to Germany, where it is far cheaper. This stone appears to the greatest advantage when set with diamonds or pearls. By candlelight it loses a part of its beauty, being apt to appear of a blackish hue. The amethyst is cut on a copper wheel with emery, and polished on tin with tripoli. This stone takes a very fine polish.

The name amethyst is from the Greek ἀμέθυστος, derived from a μεθύω, "not to inebriate,"—in allusion to the superstition that this stone had the power of dissipating drunkenness. Pliny says that the gem was so called from the fact of its approaching near to the colour

## Cairngorm, Cinnamon Stone, etc.

of wine, but not quite reaching it. In the Middle Ages, it was believed to dispel sleep, sharpen the intellect, and to be an antidote against poison. In 1652 an amethyst was worth as much as a diamond of equal weight.

## Amethyst.

Amethystine quartz or amethyst varies in color from light to clear-dark purple, sometimes nearly black, and from light to dark bluish-violet. The coloring of the stone is supposed to be due to manganese.

The best amethysts come from Brazil and Ceylon, but good specimens are found in India, Persia, Botany Bay, Transylvania, near Cork and the island of May in Ireland, at Oberstein, in Saxony, in Hungary, Siberia, Nova Scotia, Sweden, Bohemia, Canada, and in the States of Maine, Pennsylvania, Colorado, Georgia, Virginia, and Michigan.

Under heat, the amethyst turns first yellow, then green, and finally becomes colorless. The value of an amethyst depends upon the fashion, and the time has

## YELLOW QUARTZ.

been when these stones ranked among the most valuable of precious stones. At present, a fine amethyst can be bought for very little money, but should the stone become fashionable again, the best specimens will command good prices.

## THE TRIGONAL SYSTEM (HOLOSYMMETRIC CLASS)

Fig. 150 represents a simple crystal of calcite, $CaCO_3$. A first glance suggests a connection with the hexagonal system, but examination of the terminal faces reveals that the vertical axis is a triad axis only—the system is trigonal and not hexagonal. There is a centre of symmetry but no horizontal symmetry plane. Three vertical planes of symmetry meet in the triad axis, and normal to these three planes are three diad axes which we shall select as the $x\,y\,u$ directions of a Miller-Bravais axial scheme, the $z$ direction being parallel to the triad axis.

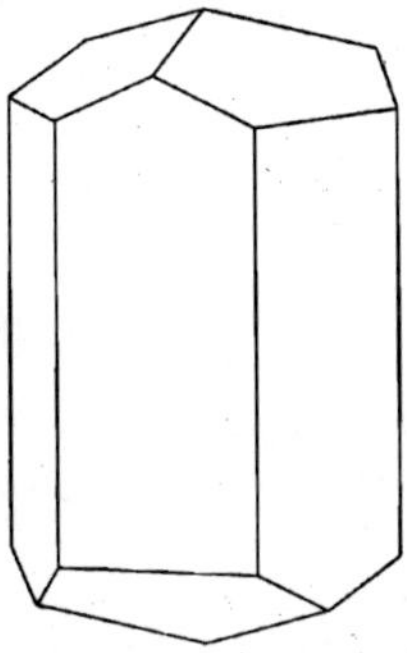

Beginning a study of the possible kinds of form, {0001} is a *basal pinacoid* as before. By inserting the poles in a stereogram (Fig. 151) and repeating them to satisfy the symmetry, the student can convince himself that {10$\bar{1}$0}

FIG. 150. A crystal of calcite.

and $\{11\bar{2}0\}$ are *hexagonal prisms*, $\{h\,k\,i\,0\}$ a family of *dihexagonal prisms*, and $\{h\,h\,\overline{2h}\,l\}$ a family of *hexagonal bipyramids* exactly as in the

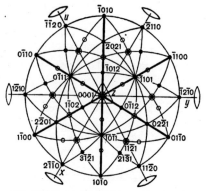

FIG. 151. Stereogram of a holosymmetric trigonal crystal.

hexagonal system. We must continue to use the term *hexagonal* for these forms, though they are special forms of a class of the *trigonal* system, since their faces are disposed regularly every sixty degrees around the triad axis. This reappearance of a number of special forms emphasises the close connection between the trigonal and hexagonal systems, and is one of the reasons why some crystallographers prefer to group all the classes of these two systems together in sub-divisions of one large hexagonal system.

A new feature arises, however, when we consider the repetition of a face $h\,0\,\bar{h}\,l$, such as $10\bar{1}1$, normal to one of the vertical planes of symmetry. The operation of the triad axis gives only three such faces, $10\bar{1}1$, $\bar{1}101$ and $0\bar{1}11$ on top, and the operation of the centre (or of the horizontal diads) gives three parallel faces below. There are no faces of this form symmetrically below the upper faces (see the separate stereogram (Fig. 152), and the form $\{10\bar{1}1\}$ is not a bipyramid but a *rhombohedron* (Fig. 153). The forms $\{h\,0\,\bar{h}\,l\}$ constitute a family of rhombo-

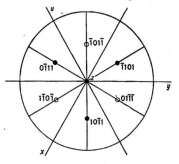

FIG. 152. Stereogram of the form $\{1011\}$ of a holosymmetric trigonal crystal.

hedra which become more and more acute (Fig. 154) as the ratio $h:l$ becomes larger. Though the face $01\bar{1}1$ is not a part of the form

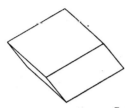

FIG. 153. Rhombohedron $\{10\bar{1}1\}$.

$\{10\bar{1}1\}$ it is a possible face on such a crystal, and its relationship to the symmetry elements shows that it, also, is repeated to give a rhombohedron (Fig. 155) geometrically similar to $\{10\bar{1}1\}$, but presenting an edge towards the observer where the latter presents a face. Developed equally together on a crystal, these two rhombohedra would simulate a hexagonal bipyramid, but even then each form would retain its own particular characteristics; in calcite, for example, $\{10\bar{1}1\}$ planes are directions of perfect cleavage.

Various nomenclatorial devices have been introduced to differentiate a rhombohedron such as $\{10\bar{1}1\}$ from its geometrically similar 'complementary' rhombohedron $\{01\bar{1}1\}$. Thus, one has been called a *positive rhombohedron* (with an upper face towards the observer) and the other a *negative rhombohedron*, but this mode of distinction seems specially undesirable in view of the established usage of + and − in optical work. *Direct* and *inverse* are more satisfactory terms, but they are not very widely used. We shall distinguish the rhombohedra by their indices; with the conventional setting of the axes, rhombohedra $\{h\,0\,\bar{h}\,l\}$ clearly present an upper face towards the observer (Figs. 153, 154), whilst rhombohedra $\{0\,k\,\bar{k}\,l\}$ present an edge in this position (Figs. 155, 156, 157}. Bearing in mind the observed simple zonal relationships of common crystal faces, we shall expect that the rhombohedra most frequently found developed together will be groups such as $\{02\bar{2}1\}$, $\{10\bar{1}1\}$ and $\{01\bar{1}2\}$

FIG. 154. An acute rhombohedron $\{h\,0\,\bar{h}\,l\}$.

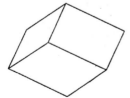

FIG. 155. Rhombohedron $\{01\bar{1}1\}$.

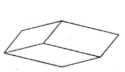

FIG. 156. An obtuse rhombohedron $\{0\,k\,\bar{k}\,l\}$

developed by successive truncation of polar edges. (The polar edges of a rhombohedron (p. 6) are those which intersect the triad axis.)

No other kind of special relationship to the symmetry elements is possible, and a face such as $21\bar{3}1$ must belong to a general form. Such a face is reflected across a plane of symmetry to give $3\bar{1}\bar{2}1$, and this

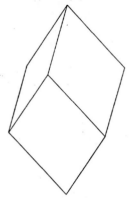

FIG. 157. An acute rhombohedron $\{0\,k\,\bar{k}\,l\}$.

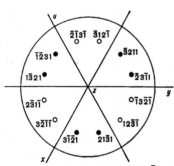

FIG. 158. Stereogram of the form $\{21\bar{3}1\}$ of a holosymmetric trigonal crystal.

pair of faces is repeated three times around the triad axis. The edges between these six upper faces, however, are not now all alike; it is a question of alternate like and unlike edges, not of six similar edges. The form is *ditrigonal* and not hexagonal. Moreover, there is no horizontal plane of symmetry, or any other element of symmetry which operates to give six faces symmetrically below; the upper faces are repeated by the centre (or by the horizontal diad axes) as shown in Fig. 158, and the form $\{21\bar{3}1\}$ is clearly not a bipyramid. Each face of the form is a scalene triangle (Fig. 159), and so the form is called a *ditrigonal scalenohedron*. The name scalenohedron is sometimes used for forms in other systems, so that we should strictly always use the full name for this particular form; when the system under discussion is clear, however, the adjective ditrigonal is often omitted. We have been considering the *scalenohedral class* of the trigonal system.

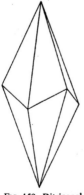

FIG. 159. Ditrigonal scalenohedron $\{h\,k\,i\,l\}$.

The list of forms for this class of symmetry now reads:

**Special forms.**  Basal pinacoid $\{0001\}$.
Hexagonal prisms $\{10\bar{1}0\}$, $\{11\bar{2}0\}$.
Dihexagonal prisms $\{h\,k\,i\,0\}$.
Rhombohedra $\{h\,0\,\bar{h}\,l\}$, $\{0\,k\,\bar{k}\,l\}$.
Hexagonal bipyramids $\{h\,h\,\overline{2h}\,l\}$.

**General forms.**  Ditrigonal scalenohedra $\{h\,k\,i\,l\}$.

Fig. 160 depicts a crystal of corundum, $Al_2O_3$, of tabular habit due to the prominent development of the basal pinacoid $\{0001\}$. The other predominant form is a rhombohedron $\{h\,0\,\bar{h}\,l\}$; in addition, there are small faces of the prism $\{11\bar{2}0\}$, a rhombohedron $\{0\,k\,\bar{k}\,l\}$ and a hexagonal bipyramid. If the predominant rhombohedron is chosen as $\{10\bar{1}1\}$, the other rhombohedron present is $\{02\bar{2}1\}$, whilst the bipyramid is $\{22\bar{4}3\}$. There are no faces of a general form present on this crystal.

A crystal of calcite of more complex

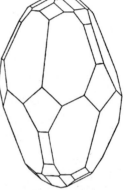

FIG. 160. A crystal of corundum.

FIG. 161. A crystal of calcite of scalenohedral habit.

development than the simple habit of Fig. 150 is portrayed in Fig. 161. The forms present are the prism $\{10\bar{1}0\}$, one rhombohedron $\{h\,0\,\bar{h}\,l\}$, three rhombohedra $\{0\,k\,\bar{k}\,l\}$ and two scalenohedra. The predominance of one of the latter confers a scalenohedral habit on the crystal.

## AN ALTERNATIVE METHOD OF INDEXING TRIGONAL CRYSTALS

Miller himself did not use a four-index notation in the hexagonal and trigonal systems, but used a three-index notation throughout. In relation to crystals with a true hexad axis his procedure was very clumsy, and it is now never used in the hexagonal system. In the trigonal system, however, his method has certain advantages over the Miller-Bravais method in some kinds of crystallographic problem and we shall explain it briefly here. His choice of crystallographic axes represents a departure from the recommendation embodied in our

## A GENERAL STUDY OF THE SEVEN CRYSTAL SYSTEMS

formulation of the Law of Rational Indices (p. 40), that 'it is convenient where possible to choose these parallel to prominent axes of symmetry', for if we follow this suggestion it is impossible to select a set of three non-coplanar axes which are symmetrically related to the triad axes. Miller therefore chose crystallographic axes parallel to the three polar edges of the *fundamental rhombohedron* (Fig. 162) (the one which we have indexed in Miller-Bravais notation $\{10\bar{1}1\}$), and thus not parallel to symmetry axes.

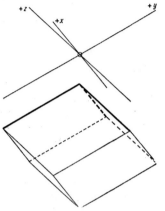

The axes are equally inclined to the triad axis and are non-orthogonal, but make equal angles with each other; this angle between the axes is the plane angle of the face of the fundamental rhombohedron (not the crystallographic interfacial angle), and it depends upon the shape of that rhombohedron in the particular substance in question. Instead of a characteristic axial ratio for each substance, we therefore have in this method of description a characteristic *axial angle* $\alpha$.

FIG. 162. Miller axes of a trigonal crystal, parallel to the edges of the fundamental rhombohedron.

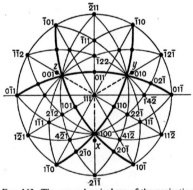

FIG. 163. The upper hemisphere of the projection in Fig. 151 indexed in Miller notation.

The crystals are still set up as before with the triad axis vertical, and are projected on a plane normal to the triad axis. Since the edges of the fundamental rhombohedron define the directions of the crystallographic axes, the indices of the three upper faces must be 100, 010 and 001 (Fig. 163). Notice, however, that the three axes do not emerge through the poles of these faces, since they are parallel to edges and not to face normals. The points of emergence of the axes $x\,y\,z$ can be located in the projection by finding the poles of the zones 010–001,

AN INTRODUCTION TO CRYSTALLOGRAPHY

001–100 and 100–010 respectively. Bearing in mind that the three axes are equally inclined to the plane of projection, we can easily determine the Miller indices of some of the forms which we have already described in the Miller-Bravais notation. Thus:

$$\{0001\} \equiv \{111\},$$
$$\{11\bar{2}0\} \equiv \{10\bar{1}\},$$
$$\{10\bar{1}0\} \equiv \{2\bar{1}\bar{1}\}.$$

The indices of further forms can be obtained by the process of adding indices in two zones (p. 53) and one example will suffice. The pole of a face of the rhombohedron complementary to $\{100\}$ lies in a zone with 111 and $11\bar{2}$, and also in a zone with 100 and $\bar{1}2\bar{1}$ (Fig. 163). Adding the first pair gives $22\bar{1}$, and since this is also in a zone with the second pair $(300 + \bar{1}2\bar{1} = 22\bar{1})$ it is the required index. The two complementary rhombohedra are thus indexed as $\{100\}$ and $\{22\bar{1}\}$ respectively, and this distinction is often a great advantage when studying trigonal crystals, in which these two forms are quite differently related to the underlying structure. (On the other hand, it was a grave disadvantage in the hexagonal system, where the adjacent faces of a single form, a hexagonal bipyramid, acquired two such different-looking indices as 100 and $22\bar{1}$; as mentioned above, this notation is now never used for truly hexagonal crystals.)

It may sometimes be necessary to convert an index $p\,q\,r$ of a face in Miller notation to the corresponding index $h\,k\,i\,l$ in Miller-Bravais notation, or *vice versa*. This is readily accomplished on a stereogram, and the faces of a number of forms on the upper hemisphere have been indexed in Fig. 163. If we adopt the convention that the particular face 100 in the one notation shall always be indexed $10\bar{1}1$ in the other, the following conversions may be useful:

$$h = p - q \quad k = q - r \quad i = r - p \quad l = p + q + r.$$
$$p = h - i + l \quad q = k - h + l \quad r = i - k + l.$$

Lightning Source UK Ltd.
Milton Keynes UK
UKOW051043120312

188798UK00001B/237/P